I0888938

Table of Contents

Rourke Educational Media
rourkeeducationalmedia.com

Can you find these words?

deer

railroad

roundabout

route

Street Signs and Symbols

Left, right. Stop, go.

Road closed. Go slow.

ROAD
CLOSED
AHEAD
SLOW

Railroad crossing. Wait here.

Speed limit. Watch for **deer**.

Curves ahead. Bike **route**.

Go around the **roundabout**.

SCHOOL
ZONE

Everywhere you want to go, there are signs that you should know.

“We’re here!”

Did you find these words?

Watch for **deer**.

Railroad crossing.

Go around the **roundabout**.

Bike **route**.

Photo Glossary

deer (deer): A plant-eating animal with hoofs and antlers.

railroad (RAYL-rohd): A double set of metal tracks for a train.

roundabout (ROUN-duh-bout): A place where traffic flows in a circle.

route (rout): A path you take to get from one place to another.

Index

About the Author

Lori Mortensen lives in Northern California with her family and her cat, Max. When she's not tapping away at her computer, she's paying close attention to the road signs so she safely gets where she wants to go, like the library, one of her favorite places.

© 2019 Rourke Educational Media

All rights reserved. No part of this book may be reproduced or utilized in any form or by any means, electronic or mechanical including photocopying, recording, or by any information storage and retrieval system without permission in writing from the publisher.

www.rourkeeducationalmedia.com

PHOTO CREDITS: Cover ©By MarkusBeck, Page 3 ©By Binksk, Page 4-5 ©Johnny Habell, Page 2,6,14,15 ©Laura Eisenberg, Page 2,7,14,15 ©HakuNellies, Page 2,8,14,15 ©By apiguide, Page 2,9,14,15 ©By Carolyn Franks, Page 10-11 ©By Tonktiti, Page 12-13 ©Tom Merton

Edited by: Keli Sipperley
Cover design by: Kathy Walsh
Interior design by: Rhea Magaro-Wallace

Library of Congress PCN Data
Street Signs and Symbols / Lori Mortensen
(I Wonder)
ISBN (hard cover)(alk. paper) 978-1-64156-186-0
ISBN (soft cover) 978-1-64156-242-3
ISBN (e-Book) 978-1-64156-293-5
Library of Congress Control Number: 2017957795

Printed in the United States of America, North Mankato, Minnesota